Rob Moore

Why Does It Fly?

and other questions about flight

WHY IS IT SO?
Science

CAMBRIDGE
UNIVERSITY PRESS

Contents

Questions about flight

Q: Why does a plane fly?

A: Four forces make it possible for a plane to fly: one to lift it up, one to pull it down, one to push it forward and one to hold it back. They are called **lift**, **gravity**, **thrust** and **drag**. Thrust and lift are the forces that keep the plane in the air. The design of planes gives them lift and they have powerful engines to give them thrust. To make the plane fly, these forces are stronger than their opposites, drag and gravity.

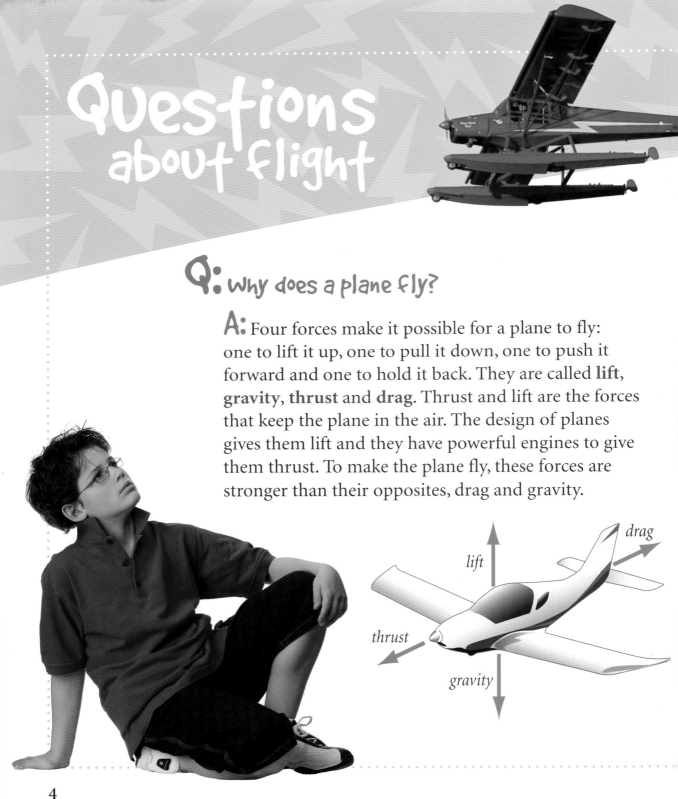

drag

lift

thrust

gravity

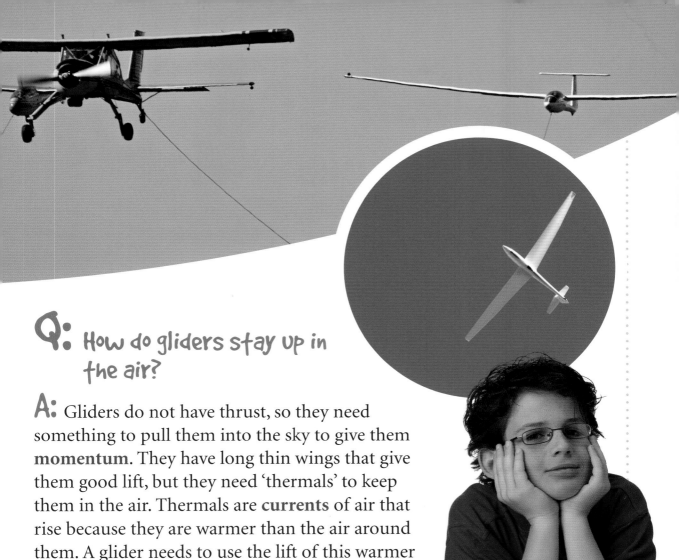

Q: How do gliders stay up in the air?

A: Gliders do not have thrust, so they need something to pull them into the sky to give them **momentum**. They have long thin wings that give them good lift, but they need 'thermals' to keep them in the air. Thermals are **currents** of air that rise because they are warmer than the air around them. A glider needs to use the lift of this warmer air to help it to go higher.

Flying high

Birds and insects use their wings to lift them off the ground and to support their weight on currents of air. People have always wanted to fly like birds, and scientists and inventors looked for ways to make wings for hundreds of years. The Italian artist and inventor, Leonardo da Vinci, drew designs for flying machines in the early 1500s.

5

Q: How do aircraft wings work?

A: An aircraft's wings are large to give it lift so it can fly. The shape of the wings makes the air on top of the wings move faster than the air below them. This makes an area of lower pressure above the wing. The air below the wing has a higher pressure, so it pushes the wing up from below, and this gives the aircraft lift. A plane that can travel very fast does not need large wings.

fast moving air over the wing = low pressure

slow moving air below the wing = higher pressure

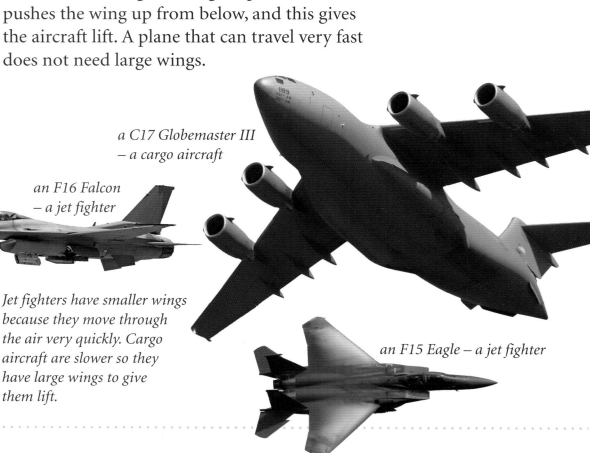

a C17 Globemaster III – a cargo aircraft

an F16 Falcon – a jet fighter

an F15 Eagle – a jet fighter

Jet fighters have smaller wings because they move through the air very quickly. Cargo aircraft are slower so they have large wings to give them lift.

Q: How does a jet engine work?

A: When you blow a balloon up and let it go, the air comes out very fast and it makes the balloon fly off in the opposite direction. A jet engine works in the same way. First it pulls air in at the front with a fan and mixes the air with fuel. An electric spark lights the mixture and the burning gases blast out of the back of the engine. This gives the plane thrust and makes it go forward.

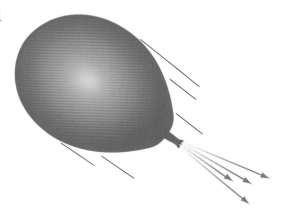

Q: Who flew the first aircraft?

A: The Wright brothers flew the first successful flying machine in 1903. It had one engine attached to the frame which turned **propellers** and made enough power to keep the aircraft moving forward at speed. The air going fast over the wings gave it lift to keep it in the air.

This is the 1903 Wright Flyer *at the Smithsonian National Air and Space Museum in Washington, DC, USA.*

Questions about bumblebees and kites

Q: How does a bumblebee fly?

A: Bumblebees have small, flat wings and round, heavy bodies so people did not understand how they could fly. In fact, a bumblebee uses its wings more like a helicopter than an aeroplane. It rotates its wings very fast, at about 200 beats a second, to get lift and flies forward by turning its wings down at the front. This turns the lift into thrust.

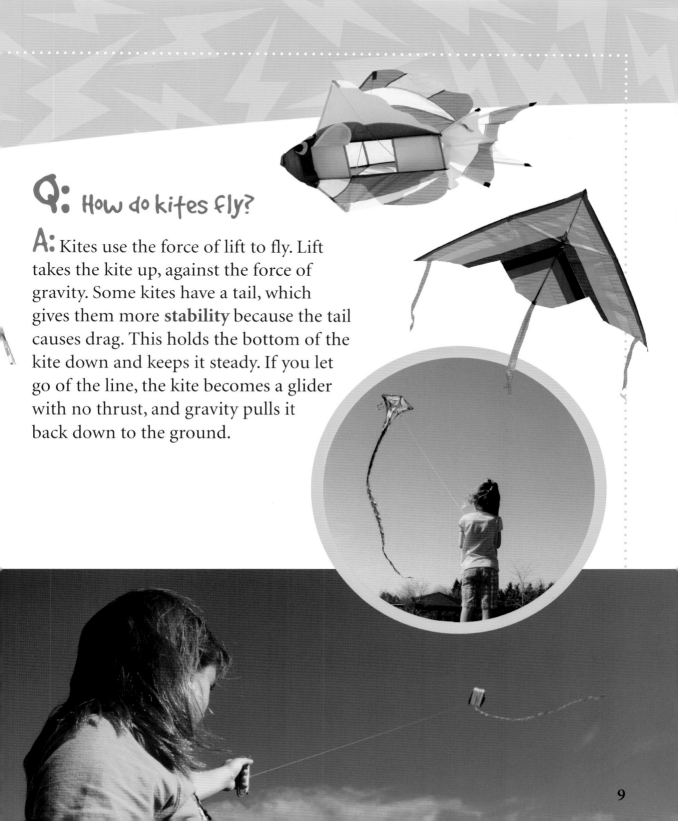

Q: How do kites fly?

A: Kites use the force of lift to fly. Lift takes the kite up, against the force of gravity. Some kites have a tail, which gives them more **stability** because the tail causes drag. This holds the bottom of the kite down and keeps it steady. If you let go of the line, the kite becomes a glider with no thrust, and gravity pulls it back down to the ground.

Questions
about space flight

Q: How does the space shuttle fly into space?

A: The space shuttle has rocket **boosters** to help it take off **vertically**. Then the rocket boosters separate from the shuttle and the shuttle goes into orbit around Earth. It returns to Earth without power, like a glider. Gravity pulls it towards the ground and its wings are shaped so it can glide down. The space shuttle uses a parachute to slow down.

The space shuttle Atlantis *takes off.*

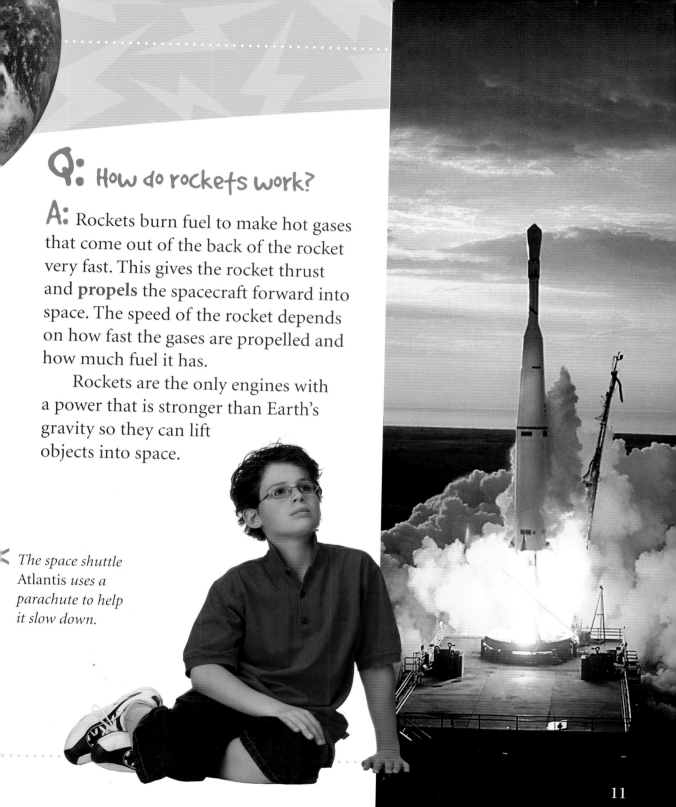

Q: How do rockets work?

A: Rockets burn fuel to make hot gases that come out of the back of the rocket very fast. This gives the rocket thrust and **propels** the spacecraft forward into space. The speed of the rocket depends on how fast the gases are propelled and how much fuel it has.

Rockets are the only engines with a power that is stronger than Earth's gravity so they can lift objects into space.

The space shuttle Atlantis uses a parachute to help it slow down.

It's a fact

> Superjumbo
The airliner which can carry the most passengers is the Airbus A380. It can carry up to 853 passengers. It is a double-deck, wide-body, four-engine airliner and its nickname is 'Superjumbo'.

> Paper plane
An American, Ken Blackburn, holds the world record for flying a paper aircraft for the longest time indoors with 27.6 seconds. He had ten throws and it was his tenth throw that broke the record.

> Jump jet
The Harrier jump jet is a fighter plane that can take off and land vertically like a helicopter, and it can also fly **horizontally** like a jet. It is like a helicopter and a jet plane in one.

> Helicopter tricks

A helicopter can go forwards, backwards, up or down, or even turn around or hover. Helicopters can therefore do things that other aircraft cannot, and they are used for rescuing people, firefighting and as air ambulances.

> Heliflight

In 1907 the French bicycle-maker Paul Cornu designed and built a powered aircraft that flew vertically. The first flight was only very short. The machine lifted its inventor about one third of a metre and the flight lasted about 20 seconds. The landing gear was made from four bicycle tyres because Cornu was a bicycle maker.

> How helicopters fly

The engine in a helicopter turns the main rotor blade on top of the helicopter. The edge of the blades is angled to force the air down as it passes over them. This pushes the blade up and creates lift. The second rotor on the helicopter helps to steer.

Flight websites for kids:

http://www.nasm.si.edu/exhibitions/gal209/wrights.htm
http://www.aviation.technomuses.ca/

Can you believe it?

Parachutes

The Italian Leonardo da Vinci drew a design for a parachute during the period 1480–1483. It was to be made of linen and a pyramid of wooden poles. Some historians think that the Chinese made the first parachutes in the twelfth century. The first parachutes were made of silk, but now they are usually made of artificial material such as Kevlar because it is stronger and lighter than silk and does not stretch too much.

Kites

The Chinese first made and used kites. In 400 BCE they used colourful kites in religious festivals. They also used them when they were at war to observe their enemies and to escape from them. Now kite flying is a popular sport.

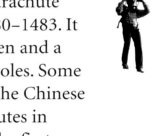

Amazing paragliding flight

In 2007 in Australia a storm carried a German **paraglider**, Ewa Wisnierska, to an **altitude** of over 9,000 metres (m) – higher than the peak of Mount Everest. At about 6,000 m ice formed on her sunglasses before she lost consciousness because there was not enough oxygen. After forty minutes she became conscious again and flew her paraglider back down. She landed safely, but when rescuers found her she was still covered in ice.

Concorde

The Concorde **supersonic** passenger aeroplane flew from Britain and France to the USA for 27 years. It travelled at about twice the speed of sound and usually took 3 1/2 hours to cross the Atlantic – half the normal flight time. Concorde stopped flying after one crashed in Paris in 2000.

Who found out?

Flight design: Leonardo da Vinci

Leonardo da Vinci (1452–1519) was born near Florence in Italy. He was a great artist and inventor who drew and designed many machines, including a tank, a submarine, a parachute and a flying machine.

He carefully studied the way the wings of birds and bats moved, and designed his flying machine so that the power came from a passenger moving his arms up and down like the wings of a bat.

First Flight: The Wright Brothers

People believe that the Wright brothers, Wilbur (1867–1912) and Orville (1871–1948), were the first people to make a machine that a person could fly in. For many years they experimented with propellers and engines.

On 17 December 1903 at Kitty Hawk, North Carolina, USA, Orville flew their aircraft, the *Wright Flyer*, for 39 minutes. It only flew about 3 m above the ground, but it made the brothers want to continue developing their aircraft.

Solo Flight: Amy Johnson

Amy Johnson (1903–1941) was born in England. She became a pilot in 1928 and in 1929 she decided to be the first woman to fly solo from England to Australia. She set off on 5 May 1930, flew 13,840 kilometres, and landed in Darwin on 24 May.

In 1931 Amy set a record for her flight across Siberia to Tokyo and in 1932 she broke the record for solo flight to Cape Town, South Africa. She mysteriously disappeared when she was flying over the River Thames on a cold and foggy day in January 1941, and no-one ever found her.

Kevlar: Stephanie Kwolek

Stephanie Kwolek (1923–) is an American scientist who invented a new, very light material called Kevlar. Kevlar is five times stronger than steel so it is very strong for its weight. Lightweight materials such as Kevlar are very important in aviation. It is used for many things such as parts of planes, safety helmets and parachutes.

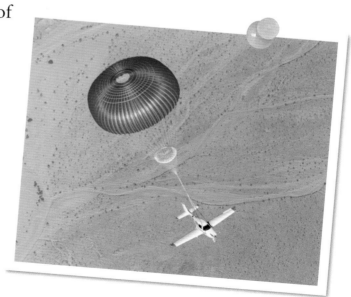

It's quiz time!

1   Unscramble the people's names.

1. oohjnsn   _Johnson_

2. vndaici   _____

3. kkeowl   _____

4. thgiwr   _____

2   Match the four forces that make it possible for a plane to fly and the names of the forces.

1. the force to lift it up   a) drag

2. the force to pull it down   b) thrust

3. the force to push it forward   c) lift

4. the force to hold it back   d) gravity

3 Choose the correct word.

1. (Helicopters / ~~Bumblebees~~ / Hovercraft) rotate their wings very quickly, at about (200 / 20 / 2,000) beats a second.

2. (Jet fighters / Rockets / Gliders) do not have thrust, so they need something to give them momentum.

3. The space shuttle has (rocket / jacket / pocket) boosters to help it take off vertically.

4. (The Airbus A380 /The Harrier jump jet /A kite) can land and take off (horizontally / vertically /at the speed of sound) like a (glider /parachute / helicopter) but flies like a jet.

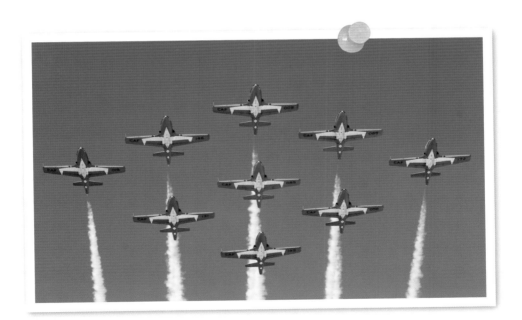

4   Complete the sentences.

1. In 1907 Paul Cornu designed and built a powered _____
    that flew vertically. The landing gear was made from four
    _____ _____.

2. Parachutes are often made of Kevlar because it is _____ and
    _____.

3. Birds and insects use their _____ to lift them off the ground
    and to support their weight on _____ of air.

4. _____ are the only engines powerful enough to overcome
    Earth's _____ and lift objects into _____.

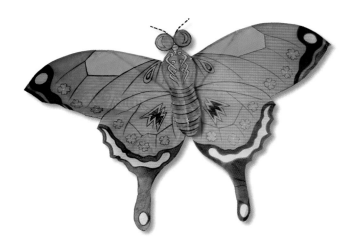

5 Find the numbers.

1. Paraglider Ewa Wisnierska was carried to an altitude of over _____9,000_____ m by a storm.

2. The record for an indoor flight by a paper aircraft is _____ seconds.

3. The 'Superjumbo' can carry _____ passengers.

4. In 1930 Amy Johnson flew _____ km from England to Australia.

Glossary

altitude: height above sea level

boosters: devices which increase power

current: movement of water, air or electricity in a particular direction

drag: the force of air that slows down a moving object

gravity: the force of attraction between objects that keeps us on Earth

horizontally: a direction parallel to the horizon, going across from left to right

lift: an upward force that goes against the force of gravity

momentum: the amount of movement in a moving object, measured by its mass and speed

paraglider: a person who jumps into the air from a great height, attached to a fabric wing

propeller: blades that spin round to drive an aircraft (or ship)

propel: move something forward very fast

stability: when something is firmly fixed, steady

supersonic: able to reach a speed faster than the speed of sound

thrust: the forward force of a jet or rocket engine

vertically: a direction at right angles to the horizon, upright